Andrey Amand

Technological aspects of functional and physiological rehabilitation

Andrey Amand

Technological aspects of functional and physiological rehabilitation

Forecasts of development of technological aspects of functional and physiological rehabilitation

ScienciaScripts

Imprint

Cover image: www.ingimage.com

This book is a translation from the original published under ISBN 978-620-7-64873-3.

Publisher:
Sciencia Scripts
is a trademark of
Dodo Books Indian Ocean Ltd. and OmniScriptum S.R.L publishing group

120 High Road, East Finchley, London, N2 9ED, United Kingdom
Str. Armeneasca 28/1, office 1, Chisinau MD-2012, Republic of Moldova, Europe
Printed at: see last page
ISBN: 978-620-7-93606-9

Andrei Amand

Technological aspects of functional and physiological rehabilitation

Predictions for the development of technological aspects of functional and physiological rehabilitation.

Keywords:

Technological Aspects; Technological Aspects Development Pathways; Predictions and Possible Development Pathways; Technological Aspects of Functional Rehabilitation; Technological Aspects of Physiological Rehabilitation; Virtual World; Coding Basics;

Annotation:

Information technology is redrawing the map of modern life, influencing its many aspects and actively contributing to the emergence of innovative approaches in every field, from education to health care. Particularly significant is their contribution to the development of functional and physiological rehabilitation, where the synergy between classic technological developments and the latest innovations opens up limitless possibilities for improving quality of life and accelerating recovery processes.

In this context, optical storage media with advanced security technologies become a valuable tool for storing and transmitting knowledge and rehabilitation programmes. They provide a secure means of storing educational materials, relaxation videos, training and simulations, while meeting the high requirements for the protection of personal and medical data.

The book you hold in your hands offers a comprehensive look at the progressive technological aspects of functional and physiological rehabilitation, including the use of VR and AI. By highlighting the role of innovative protective mechanisms of optical media, we aim to show how these advanced technologies are transforming rehabilitation approaches, providing safer, more accessible and personalised treatment experiences.

At the intersection of these constantly evolving technologies, a new understanding of integrative technical solutions and their implications for future generations of professionals and patients striving to achieve a fulfilling life is emerging. The exploration of new boundaries between physical reality and virtual space holds both challenges and unlimited prospects for the inventions that our publication explores in the context of rehabilitation and technological defence.

Content:

TECHNOLOGICAL DEVELOPMENT FORECAST UNTIL 2030

The US President's annual report and address to Congress paid special attention, and one could even say the main attention, to the issues of innovative modernisation. The address touched upon several specific areas of innovative development, which in our previous forecast were highlighted as the latest basic technologies. One of the decisive directions was the full implementation of high-speed internet in all corners and districts of the United States, regardless of their remoteness and not always favourable weather conditions.

In our forecast, on the basis of analysis of published reports on research and development, on the basis of system analysis of the patent and licence situation and trends of its development and preconditions for fundamental and local changes, we selectively selected four groups of fundamental technologies that will determine the nature and ways of development of mankind in the 21st century:

THE LATEST BASIC TECHNOLOGIES - REVOLUTIONARY TECHNOLOGIES, ON THE BASIS OF THE APPLICATION OF WHICH THE DEVELOPMENT OF SOCIETY AND ITS FORMATIVE FOUNDATIONS CAN BE FUNDAMENTALLY CHANGED

1. ELECTRONICS AND COMPUTER SCIENCE

1.1. Microelectronics;

1.2. Terabit memory;

1.3. Superconducting devices;

1.4. Super smart chips;

1.5. Self-replicating chips;

1.6. Optical electronics, terabit optical memory devices, terabit optical communication devices, elements and nodes of optical computers and control systems of different levels based on optical terabit memory.

Undoubtedly, the widespread adoption of high-speed internet has heralded dramatic changes in information sharing and data management. However, with all the benefits that fast digital communication brings, the vulnerability to potential cyber threats is also increasing. Undoubtedly, information security is one of the key challenges of our time, necessitating the development and implementation of innovative defence technologies.

In an era where global cyberspace is becoming the scene of unprecedented cyber warfare, ensuring the reliability and security of data carriers is of paramount importance. The development and introduction of viruses capable of inflicting damage on the level of military attacks requires a response that matches the growing complexity and sophistication of these threats.

Modern information security solutions range from simple encryption mechanisms to complex multi-layered security systems, including the application of artificial intelligence mechanisms to predict and neutralise cyberattacks, as well as the development of predictive algorithms to automatically identify and manage vulnerabilities in the network.

Identifying, analysing and comparing different approaches to information security, as well as characterising the latest innovative technologies in this field, is a crucial task. My efforts are aimed at finding optimal solutions that will ensure the safe implementation of high-speed Internet without creating additional risks in an environment where the novelty and complexity of cyber threats await at every turn.

To defend against cyberattacks, the network uses layered strategies and technologies tailored to different threats and security requirements. Here are some of the key measures to counter cyber attacks:

- Antivirus software: provides basic protection against known viruses, Trojans and other malware.
- Firewalls: control incoming and outgoing traffic by blocking unauthorised access to network resources.
- Data encryption: protects transmitted and stored information by making it unreadable without the appropriate decryption key.
- Intrusion Detection and Prevention Systems (IDS/IPS): analyse network traffic for suspicious activity and block potential threats.
- Access control and authentication: allows you to control who can access certain data and resources and how.
- Data backup: regular backups of important data helps to restore information in case of an attack.
- Security updates and patches: address known vulnerabilities in software and operating systems.
- User education: raising users' awareness of cyber threats and developing skills for safe online behaviour.
- Network segmentation: dividing the network infrastructure into parts to limit the propagation of an attack within the network.
- Monitoring and analytics: regularly analyse and monitor network activity in real time to detect and respond to attacks.
- Using Virtual Private Networks (VPNs): hiding network activity and protecting data in transit over public networks.
- Multi-factor authentication: requires two or more proofs of user legitimacy to access resources.
- Security policies and standards: develop and implement internal policies and procedures to ensure information security.

Applying these measures in combination can significantly reduce the risk of successful cyber attacks and minimise the potential harm from them.

A security technology for optical discs based on the principles of dimensional identification of marking conductive layers offers a solution to a number of existing problems in the market for optical storage media systems:

- Improved security: unlike existing security systems that are easily hackable, this technology offers a unique identifier that is difficult to decipher, significantly increasing the level of protection against unauthorised copying and seizure of information.

- Efficient identification and control: through the introduction of conductive layers with unique characteristics, optical discs can be quickly and efficiently identified by their ownership. This allows media to be accounted for and controlled within the corporation, preventing unauthorised introduction of new media and removal of existing media.

- Information leakage prevention: This technology can be used to restrict the use of storage media exclusively within the corporate network. By communicating with the drive servos, the marking layers can block reading and writing of data if the drive is attempted to be used outside of its authorised location, thus avoiding information leakage.

- Compatible with personal computers without hard drives: the specialised coding of optical discs allows them to be used as a means of user identification, providing access to corporate resources and software products via the Internet. This makes them a convenient tool for cloud- and network-oriented portable devices.

Thus, the principles of dimensional identification and the use of marking conductive layers represent a powerful tool in combating a number of key information security challenges in today's corporate world, providing an effective solution for information security at the physical level.

There are many variants of thicknesses of coding coatings, which allow to have many variants of protective code, unlike the known technologies, which have only one variant of code. In the process of coating is applied control technology is fully identical to the decoding technology, which allows full control of the quality of coding in the process of manufacturing discs, without removing the disc from the conveyor belt, unlike existing technologies, in which the disc for control must be removed from the conveyor belt and installed in the control device, thus control selective, and in the proposed technology - 100% control, which excludes the release of defective discs, which in existing technologies are detected only in the existing technologies In the proposed technology it is possible to encode all categories and types of discs regardless of the format of recording and reading, in contrast to existing technologies, in which encoding depends on the format of recording and reading disc. The encoding coating can provide the basis for a personal secret code or cipher, which is not available in existing technologies. The decoding and identification sensor is mobile and can have several variants of delivery, including a stand-alone variant, not connected with the disc drive, and in existing technologies the decoding system is installed only in the disc drives, thus, it is possible to control the presence and correctness of coding only in the process of installing the disc in the disc drive, and in the proposed technology to control and identify the code can be outside the disc drive, for example, in shops or at the entrance of enterprises and institutions, which is especially important to ensure the Decoding eliminates any dependence on the optical systems of the disc drive, but the results of decoding can change the operation of optical systems, such as servo - drive for orientation and control the position of the focus of the reading or writing laser, in contrast to existing technologies in which the decoding process depends entirely on the optical elements of the disc drive, which complicates its design and dramatically reduces reliability. The proposed technology has several hierarchies of the principal working scheme, has a flexible algorithm and can be embedded in any optical memory security system, including hybrid storage media, having besides the optical component and media built on other basic principles; existing technologies do not have this flexibility; allows you to use the disc code as an input password to enter the professional information arrays of the Internet, which is not available in existing technologies.

Preliminary definition of the basic product of the project:

- The main sector of the market is corporate clients:

Banks and finance companies;

Industrial Corporations;

Scientific and research laboratories;

Transport companies, railway stations, airports and seaports;

Major retailers;

Municipal Services;

Government organisations and institutions;

Major medical facilities;

Insurance companies;

The branches of the armed forces;

Police and intelligence services.

- Preliminary assessment of market size:

- The average price of a disc drive with an integrated magnetic resonance sensor is $350;
- The average price of a disc with a coding ring is $5;
- The average number of disc drives required for one corporate customer is 1750; the price of the disc drives for said corporate customer is $612500;
- The average number of discs required for one corporate customer is 17500 - 10 discs per disc drive; the price of discs for said corporate customer is $87500;
- The approximate number of corporate clients is more than 10,000;
- The approximate market size for corporate customers is: $70,000,000 per year;
- In total, across all categories of corporate clients, the estimated annual market size could be more than US$700,000,000,000 in the first phase.

- Product Characteristics:

The project product is an optical disc with an encoding ring and a disc drive with an integrated sensor module, usually consisting of three micro sensors. If necessary, the sensor module can be supplied without the disc drive. If required, the project company can also provide services for corporate customers by organising the commissioning of the encryption and data protection system on a turnkey basis.

Principle bases of protective coding of optical media or information storage devices, mainly in the form of a disc, transparent to the light flux coming from the output optical system of a single-mode laser diode, having standard executive dimensions: outer diameter of 120 millimetres, and thickness of 1.2 millimetres; the disc is glued from two halves, each with thickness of 0.6 millimetres; the coating is applied on one of the halves of the disc on a ring with outer diameter of 120 millimetres and inner diameter of 118 millimetres; the thickness of the coating is 0.6 millimetres.

The conceptual basis of coding consists in the following principle: the coding signal is formed from the reaction of a sensor or a group of sensors to the thickness of the annular coating on the disc, comparison of the received signal with a statistical standard of this signal, the equivalent of the resonant reaction of sensors to the thickness of the coating, specific indicators of the coating material, conductivity of the coating material, density of the coating material, electrical resistance of the coating material. In the system of servo marking of the formatted disc, which, as a rule, has the form of group combinations of servo - points on the information tracks of the disc, instead of one of the points of the group combination, the signal from the decoding sensor of the protective coding system is inputted, and, in case of coincidence of the integrated signal from three sensors with the specified signal parameters, the servo system of the disc drive starts to orient the laser focus on the information track, and, thus, the system starts the process of reading or writing on the optical disc. In case of mismatch between the signal from the sensors and the statistical form of the signal in the disc drive processor memory, the disc drive servo system does not orient and stabilise the laser diode beam focus path on the disc information track and reading or writing on the disc becomes impossible.

Identification of the disc in the disc drive can be carried out by measuring the coating thickness in real time, comparing the measurement results with the statistical value of this parameter stored in the disc drive processor and outputting a signal to the comparison device in the disc drive processor. The identification process may be carried out when the disc is rotated or when the disc is inserted into the disc drive. During identification when the disc is inserted into the disc drive, negative identification results do not allow the inclusion of any disc drive structure, and vice versa, a positive identification signal enables the necessary disc drive structures.

Elements of the protective system of resonance coding-decoding can be built into any existing disc drive design implementing all known optical memory technologies without any design or circuit limitations. Existing disc drives can also be modified for installation of a microsensor system by cutting a sensor micromodule into the supporting structure of the disc drive housing. If necessary, coating can be performed on existing discs.

The manufacture of an optical disc with a protective coding coating does not require special technologies and equipment. Modernised technological equipment currently in use can be used for manufacturing. Application of coding coating can be combined with making a copy of the disc in a mould using a master disc with an identification point in a formatted servo-marking system, which thus will be imprinted on each information track, and there are more than 37000 of them in a conventional optical disc.

The approximate scheme of using discs with protective coding-decoding at corporate clients provides for manufacturing for each such client a certain number of discs with thickness parameters and coordinates of microsensors inherent only for this client. The design and technical characteristics of the sensor micromodule can also be upgraded based on the customer's wishes, but in accordance with the control parameters of the protective coding coating on the discs.

Discs with security coding can be used in Blu-ray and HD DVD systems, besides, the security coding system can be applied in new developments and technologies of optical digital memory, including discs with particularly high recording density, multilayer discs, monolithic optical discs with memory capacity of 1 terabit or more.

When manufacturing discs, the necessary indication in the servo marking can be made during the pressing process; the servo drive of the disc drive starts the orientation of the focal point of the laser beam only when the coding signal from the coding and decoding system, formed by a system of three micro-sensors, which by means of magnetic resonance methods, compare the thickness of the coating with the standard and if the signal parameters match the standard at least two sensors, add the received signal to the system of symbols and marking points of the servo-marker.

The technology of manufacturing discs for personal computers is similar to the technology of manufacturing such discs for other variants of optical memory. The technique of using discs with security coding is formed based on the type of computer, the degree of its saturation and power, speed, etc. Of particular importance is the possibility of using the technique and technology of security coding in hybrid discs being created, combining a hard disc with an optical disc.

Now back to the earlier prediction:

1.7. Bioelectronics, including biosensors; bio - computers;

1.8. Information systems hardware, including super - parallel computers; neuro - computers;

1.9. Software, automatic translation systems; reality modelling systems (VIRTUAL REALITY SYSTEMS); self-replenishing databases.

2. NEW MATERIALS

2.1. Ceramics, including superconductors (coils possessing the property of superconductivity at high temperatures); super heat conductors - nanocomposites based on artificial and natural diamonds; gas turbines and engines created using ceramic materials; new types of glass (nonlinear optical glass); new types of coatings on glass and ceramics, significantly changing their properties;

2.2. Semiconductors, including optical integrated circuits; superlattice semiconductor elements;

2.3. Metals, including amorphous alloys; alloys with absorbed hydrogen; magnetic materials;

2.4. Organic materials organic nonlinear optoelectronic elements; memory based on optical hole burning; molecular devices; thermoplastic molecular composites;

2.5. Composite materials, high quality carbon fibre reinforced plastics; high quality metal composites; high quality ceramic composites; high quality carbon-carbon composites (carbon-carbon, modified graphite, pyrolysed graphite, multi-stage pyrolysed graphite, electrochemically activated graphite, on flexible or elastic viscose backing with subsequent electrochemical activation after application of pyrolysed graphite to the viscose matrix).

3. SCIENCE OF LIFE

3.1. New types of medical drugs, including drugs for the treatment (prevention) of tumour diseases; drugs for the treatment (prevention) of senile dementia; drugs for the treatment (prevention) of diseases of the immune system and allergies;

3.2. Utilisation of human somatic features, including bone marrow bank; bioenergy;

3.3. Production of artificial bio-objects, including artificial organs; artificial enzymes and membranes.

BASIC TECHNOLOGIES THAT SUPPORT PRODUCTION ACTIVITIES, ENSURING THE COMPETITIVENESS OF THE INDUSTRY ON THE WORLD MARKET

4. ENERGY

4.1. Energy production technologies, including fuel cells; solar energy sources; alternative petrol-water emulsions; small-sized light water reactors with their own stability; nuclear fusion reactors; high-speed multiplier reactors;

4.2. Energy efficiency technologies, including high-efficiency refrigeration and heat pumps; superconducting energy condensers.

5. AUTOMATISATION

5.1. Robotisation, robots with artificial intelligence; micro-object handling devices;

5.2. Technologies in the field of machining equipment, machine tools with artificial intelligence and computer numerical control; complex machining centres; ultra-precision machining machines;

5.3. CAD/CAM technologies, computer-aided design and manufacturing, computer-aided design systems with artificial intelligence; product modelling;

5.4. CIM/HIM technologies (complex-integrated and highly integrated manufacturing), autonomous systems with distributed control; integrated process equipment.

SOCIALLY IMPORTANT BASIC TECHNOLOGIES THAT HELP RAISE LIVING STANDARDS

6. LIAISON

6.1. Satellite and mobile communications systems, including personal communications; data networks based on ultra-small earth stations (VSAT) and satellites;

6.2. Image transmission, including high definition television (HDTV); cable television systems for satellite communications - radio programme transmission (CS/DC-CATV);

6.3. Multichannel communications, television conferencing systems; video telephones;

6.4. Communications network development, broadband integrated digital communication network (ISD) switches; optical subscriber communication systems; local optical communication networks.

7. TRANSPORT

7.1. Railway transport, means of transport with linear motor based on the principle of superconductivity. New generation of vehicles with linear motor driven by the principle of superconductivity at high temperatures; high-speed land transport with linear motor (HSST); advanced train control system (ATCS); bimodal systems (end-to-end traffic system);

7.2. Car production technology, new generation cars (with combined engines, with engines running on emulsions of petrol and water or solar oil and water); cars with alternative energy sources (electric cars); revolutionary car production technologies;

7.3. Shipbuilding, techno-superliners; surface gliding vessels; artificial intelligence vessels; aquarobots;

7.4. Air transport, multi-seat passenger aircraft; hypersonic transport aircraft; small propeller aircraft with vertical take-off and landing.

8. SPACE UTILISATION

8.1. Space exploration technologies, underground facilities for weightlessness experiments; research bases on the surface of the moon; linear propulsion catapult;

8.2. Ground technologies, construction of super skyscrapers; super large air domes; technologies for dismantling super skyscrapers;

8.3. Utilisation of underground space, underground freight flow networks; construction of underground highways and railways at great depths; underground heat condensation systems;

8.4. Ocean utilisation, creation of artificial islands; floating stations; marine grazing grounds; marine recreational areas.

TECHNOLOGIES AIMED AT COMBATING ENVIRONMENTAL DEGRADATION, CREATING RESOURCE-SAVING PRODUCTION FACILITIES, AND CREATING WASTE-FREE PRODUCTION FACILITIES

9. ECOLOGY

9.1. Measures related to the general warming of the earth (climate warming), CO2 sequestration technologies using a catalyst; CO2 sequestration technologies using plants; CO2 sequestration and recycling technologies;

9.2. Combating ozone depletion, Freon replacement gases; Freon regeneration technologies;

9.3. Waste control, self-destructive plastics; underground systems for treating conventional waste; underground water storage and treatment facilities;

9.4. Technologies of resource-saving productions;

9.5. Waste-free technologies;

9.6. Technologies for decontamination of contaminated objects;

9.7. Technologies for cleaning water environments from radioactive contamination;

9.8. Technologies of personal protective equipment against man-made and natural disasters;

9.9. Technologies of personal protective equipment against terrorist acts;

9.10. Technologies of disposable personal protective equipment against man-made and natural disasters;

9.11. Technologies for disposable personal protective equipment against terrorist attacks;

9.12. Technologies to protect individual dwellings from man-made disasters;

9.13. Technologies to protect individual dwellings from natural disasters;

9.14. Alternative energy technologies for individual dwellings in case of emergencies;

9.15. Technologies for producing synthetic water from air.

Optical media with advanced protective technologies have a wide range of applications in various fields, including technological aspects of functional and physiological rehabilitation. Here are some examples of their practical use:

- Patient education programmes: Optical discs can store educational videos and programmes aimed at recovery from injury or surgery. These programmes, including physical therapy techniques, meditation and breathing exercises, can be used by patients at home to speed up the rehabilitation process.

- Relaxation and meditation suites: specialised compositions and videos for relaxation, meditation and stress management can be placed on optical media, making them easily accessible to patients. These packages can be used both in rehabilitation centres and at home.

- Training and training programmes for rehabilitation: optical discs containing motor recovery and training programmes with a specific set of exercises and instructions on how to perform them can be an important element of physical rehabilitation.

- Physiotherapy and occupational therapy programmes: discs can serve as a carrier for specialised programmes designed to stimulate motor skills, coordination and regain self-care skills, which are also important for successful rehabilitation.

- Use in psychotherapy and psychological care: optical discs can contain materials for psychotherapy sessions, including visualisations, music and anti-anxiety programmes, allowing patients to use them for self-help and as an adjunct to mainstream treatment.

- Cognitive impairment programmes: optical media can host cognitive training and programmes designed to stimulate memory, attention, logical thinking and other cognitive functions for a variety of impairments.

Advanced security technologies for optical media ensure that this information is protected from unauthorised copying, preserving copyright and content confidentiality, which is critical for medical and rehabilitation information.

Further examples of practical use of optical media and aspects of functional and physiological rehabilitation in previously implemented projects are given.

INFORMATION STELE

The information pedestal can be integrated with an optical media system to enhance its ability to provide and protect information. The relationship between the information stele and an optical disc with an encoding ring and a disc drive with an integrated sensor module can be used for the following purposes:

1. Improved information security: The encoding ring on the optical disc and the sensor module in the disc drive can be used to create a highly efficient system for authenticating and encrypting the data provided by the information pedestal. This provides an additional layer of protection for sensitive information.
2. Interactive user interaction: the touch module can monitor and react to user interaction with the information wall by activating the viewing of specific information or launching multimedia content stored on the optical disc.
3. Personalisation of information: by using sensors to identify the user or their preferences, the system can provide personalised information, improving the user experience.
4. Turnkey implementation of the information coding and protection system: comprehensive services include not only the installation of the stele and integration with optical media, but also the creation of specialised software to ensure the effective operation of the entire system.

Application Examples:

1. Educational institutions: information steles can be used in universities, schools and museums to provide educational content, interactive maps and instructions.
2. Shopping and business centres: to provide information on shop locations, promotions, events and services with the added protection of confidential customer and partner information.
3. Public institutions and organisations: in office buildings and exhibitions, information pedestals can be a means of raising visitor awareness and improving visitor interaction.

The integration of an information pedestal with an optical media system and a sensor module increases the functionality of the pedestals, ensures data security and creates innovative opportunities for user interaction in various projects.

There are several key reasons and advantages to using an encoding ring on the optical disc of an information stele:

1. Data protection: The encryption ring increases security by encrypting the data stored on the optical disc. This prevents unauthorised access and copying of information.
2. Authentication: the encoding ring may contain unique identifiers that allow verification of the authenticity of the disc and the data provided, thereby protecting against counterfeiting.
3. Access control: the system can be configured to read and process information only from discs that have the correct coding. This restricts the use of the pedestal to authorised media only.
4. Software integration: specialised stele software can use the encoding ring to dynamically load and display content, thereby allowing information to be easily updated and the functionality of the device to be extended.

The use of an encoding ring on the optical disc of the information pedestal reinforces the principles of information security, data distribution efficiency and content management flexibility.

The use of an encoding ring on the data stack plays a key role in managing and controlling access to data:

1. Data encryption: a ring can carry the information needed to decode or decrypt the content on an optical disc, so without the correct encoding ring, the data is inaccessible.

2. Unique physical marking: as a physical security mechanism, the encoding ring on the disc serves as an additional layer of security that is required for authentication before interacting with the content.

3. Access authorisation: only disc drives or optical readers with the appropriate sensor module are able to read information from the ring, restricting access to authorised devices only.

4. Software interaction: the information pedestal software can be configured to perform actions only after successful authorisation of the encoding ring, thus ensuring that only the correct optical disc is used.

5. Preventive measures against misuse: ring coding acts as a preventive mechanism that prevents information from being copied or modified without authorisation.

As a result, through the use of coding rings on information pedestals, access to data can be effectively controlled, minimising the risk of unauthorised use or disclosure of confidential information.

To ensure the physical security of information steles, apart from the use of encoding rings on optical media, various methods and devices can be used:

1. Vandal resistant enclosure: robust and resistant enclosures prevent physical damage to the pedestal and the devices mounted in it.
2. Locks and locking mechanisms: locks on media compartments and other critical components provide protection against unauthorised access.
3. Electronic means of protection: use of electronic locks and alarms activated by unauthorised access.
4. Weather protection: waterproof and UV resistant materials protect the pedestal from rain, snow and direct sunlight.
5. Temperature control: integrated cooling and heating systems prevent overheating or overcooling of important components of the pedestal.
6. Ultrasonic or infrared sensors: to detect and warn of unauthorised approach to the pedestal.
7. Access control systems: biometric systems such as fingerprint scanners or facial recognition can be used to control access to stele functions.
8. Base anchoring: securely attaching the pedestal to the floor or other surface to prevent it from being moved or stolen.
9. Screen protectors: scratch and impact resistant screen protectors keep your screen intact and readable.

The use of these methods of protection can significantly reduce the risk of vandalism, theft, weather damage and other external factors, thereby increasing the longevity and reliability of information steles in various conditions.

Ultrasonic and infrared sensors play an important role in the system of protecting information pedestals against unauthorised access or vandalism: These sensors can monitor objects that approach the pedestal. If an object comes too close or tries to interact with the stele unauthorisedly, the sensors are triggered and activate a specific response system - this can be an audio or visual warning, notification to the security service, blocking of stele functions, etc.

If unauthorised access to the information pedestal occurs, various types of warnings can be used to ensure safety and equipment protection:

1. Audio - Warnings: the loudspeakers can trigger warnings or messages such as "Please move away from the unit" or "Unauthorised access is prohibited".
2. Visual Warnings: the stele screen can display colourful flashes or specific graphic warning messages.
3. Notifications to guards: the security system can automatically send notifications to the guard post or security units.

What are smart city solutions?

Smart city" (smart city in English) is a concept of integrating several information and communication technologies and the Internet to manage urban infrastructure.

The complex includes:

- interactive information panel;
- two immersion geomagnetic observation devices.

The information panel is additionally equipped with:

- two outdoor surveillance cameras monitoring the space in front of the panel;
-

three microphones and acoustic systems to form areas of comfortable interaction between three users at the same time.

RealTime face motion capture technology

The use of this technology makes it possible to organise communication between residents and responsible persons of a district or city, represented as their avatars, controlled by specially trained personnel. The control takes place in real time without any specialised hardware. The avatar's emotions are controlled by the operator; its body position and behaviour correspond to predetermined emotional patterns chosen by the operator in the process of communication.

IMMERSIVE GEOMAGNETIC OBSERVATION DEVICES

Immersive geomagnetic observation devices play a significant role in creating solutions for smart cities, enabling interactive user interaction and increasing the efficiency of urban services:

Relaxation and rehabilitation:

Immersive geomagnetic viewing devices can be used in clinics and rehabilitation centres to reduce stress and anxiousness in patients by providing calming visual scenarios that promote relaxation and mental recovery.

Scientific Application:

In academic environments and research centres, such devices can be used to visualise complex data and models, providing a deeper understanding of research problems and results.

Protection against Unauthorised Access:

Integration with the optical media system allows the information provided via immersive devices to be protected from unauthorised access, ensuring data control and protection.

Smart City applications:

Immersive geomagnetic surveillance devices can be used in urban information systems, transport, medical facilities, energy and waste management, and law enforcement agencies to improve the quality of life of citizens and optimise urban infrastructure.

The use of immersive technologies, combined with data from city residents and electronic devices, can increase the efficiency of urban management, improve the environment and promote socio-economic development. Systems based on artificial intelligence, big data and predictive analytics complement the functionality of such devices, turning cities into smart ecosystems where information serves as a basis for strategising and decision-making in various spheres of urban life.

An optical media system can integrate with immersive devices in a way that protects and manages information. Here is how it can be realised:

1. Data encryption: Information on optical media can be encrypted using advanced encryption algorithms. This ensures that even if the media is physically stolen, the data remains protected from unauthorised access.

2. Authentication of data access: An optical media system may require authentication before granting access to the data. This can be implemented through special encoding rings or other identification mechanisms that must be recognised by the immersive device.

3. Cloud integration: Optical media can be used to store encryption keys or access tokens to cloud systems where the underlying information is stored. This provides another layer of protection, as data can only be accessed with physical media.

4. Temporary access restrictions: Access to information via immersive devices may be restricted by time limits. Data on optical media may be automatically blocked or deleted after a certain time has elapsed.

5. Physical protection of storage media: optical media can be installed in enclosures that are protected against impact and vandalism. Such enclosures can be integrated directly into immersive devices, minimising the risk of unauthorised physical access.

6. Access logging and monitoring: every time an optical media is accessed via the immersive device, this fact can be logged. This ensures that all access attempts can be tracked and analysed if necessary.

This integration not only effectively protects valuable information, but also enhances the ability to deliver and manage it.

The following videos can be used to achieve relaxation and rehabilitation with immersive geomagnetic observation devices:

- Nature scenes: videos showing beautiful natural scenery such as woodlands, mountain rivers, ocean shores or tranquil lakes encourage relaxation and can be used in stress therapy.

- Views of space: videos showing the grandeur and scale of space can enhance the sense of relaxation and provide a unique opportunity for space travel without physical movement.

- Meditative and abstract images: slow motion and calming footage, abstract visual patterns or meditative animations can promote mental relaxation and attention.

- Underwater World: immersive videos depicting underwater scenes allow you to immerse yourself in the world of marine animals and reefs, bringing peace and positive emotions.

- Scenes from the past: reconstructions of historical events or ancient worlds can be used for cognitive rehabilitation and to stimulate interest in history and culture.

Immersive geomagnetic observation devices can be used in the following areas of science:

1. Astronomy: Studying space objects and phenomena using immersive videos simplifies the explanation of complex concepts and improves the comprehension of astronomical data.

2. Biology and Ecology: Using videos of wildlife and ecosystems helps with education and learning about conservation.

3. Medicine and psychotherapy: Immersive videos can be used for meditation, relaxation and as part of art therapy.

4. Education and training: Immersive content videos enhance the learning process by providing more vivid visualisations for a better understanding of complex topics.

5. Archaeology and History: Reconstructions of historical scenes and monuments can be useful for the visual study of history.

6. Geography and Cartography: Creating interactive maps and landscapes helps to better understand geographic data and the study of geology.

Immersive technologies accelerate the progress of research, enhance educational experiences, and advance various fields of science by creating immersive visual and interactive environments.

Immersive geomagnetic monitoring devices can be used to improve sleep and reduce anxiety through the following principles:

1. Creating a Calming Atmosphere: Immersive videos with calming nature scenes or slow-moving abstract images can create the relaxing environment needed to fall asleep quickly and sleep deeply.

2. Meditation and Mindfulness: Videos that include elements of meditation can help viewers focus on the present moment, which helps reduce anxiety levels.

3. Noise Suppression: Immersive devices can be used to play environmental sound (e.g. forest noise, wave noise, white noise) which has a calming effect and helps to divert attention from restless thoughts before bedtime.

4. Positive Visual Stimuli: Using visual stimuli with positive images and stories can help reduce anxiety and create a positive mood before bedtime.

The use of immersive technologies at home, as part of a bedtime ritual or in therapeutic centres for relaxation sessions, can facilitate the management of insomnia and anxiety disorders. When properly integrated with stress management strategies and cognitive behavioural therapy, immersive devices can enhance the effects of this method in holistic therapy.

The frequency of use of immersive geomagnetic monitoring devices to improve sleep can vary depending on the individual needs and preferences of the user, as well as the recommendations of experts. Recommendations are usually based on the following principles:

1. In the initial stages: You can start with frequent use, such as daily for the first few weeks, to improve sleep quality and form healthy sleeping habits.

2. Incorporating it into your bedtime ritual: Regular use of the device as part of your bedtime ritual can promote positive sleep reflexes - for example, watching immersive videos 30 to 60 minutes before bedtime.

3. As needed: If the purpose of using the device is to reduce anxiety, it can be used as needed, for example, during periods of high stress or before anxiety-provoking events.

4. Following medical advice: If the use of the device is part of a therapy programme, it is important to follow the instructions of a specialist who can individually determine the optimal frequency and duration of sessions.

It is important to monitor the body's response to immersive sessions and, if necessary, adjust their frequency and duration to avoid the effects of habituation or stimulus overload before bedtime. The optimal approach to the use of immersive devices is moderation and combination with other methods to improve sleep quality, such as relaxation exercises, sleep hygiene and avoidance of stimulants before bedtime.

VIRTUAL SIMULATORS, VIRTUAL SIMULATORS VR TRAINING

In this age of rapid technological discovery, virtual reality (VR) continues to expand beyond the gaming industry to become a powerful tool in training and rehabilitation. Virtual VR simulators, embodied through state-of-the-art visualisation systems actively used in scientific research, are finding applications in a variety of industrial and medical fields. These tools are overturning traditional training methods, transforming them from costly and time-consuming activities into an effective and cost-efficient means of developing professional skills. Optical media play an important role in the deployment of VR technologies as they provide a convenient and reliable means of storing the extensive data required for a full volume immersive experience. The encryption capabilities of optical discs such as CD, DVD and Blu-ray provide an additional layer of protection against both unauthorised access and the ability to copy content. The use of sophisticated encoding and identification techniques, including special conductive layers, significantly increases the security of stored data and ensures its integrity. VR training applications often require big data, resource-intensive visualisation and scaling scenarios that network traffic may not be ready for. Optical media is becoming the answer to such requirements, allowing users to access the necessary programmes and simulations even when internet connectivity is limited.

With these trends in mind, VR combined with secure optical media holds great promise for the field of functional and physiological rehabilitation, providing educational and therapeutic programmes accessible to anyone in need of effective and personalised rehabilitation techniques. Compatibility, security and ease of use make this combination of technologies the basis for the future of medical training and therapy.

Виртуальная копия

Реальный объект

DEVELOPMENT OF VIRTUAL REALITY IN THE MODERN WORLD

Involvement in the virtual world, using special applications that simulate reality as much as possible, allows a person to choose the optimal solution to a critical situation. Not all devices that are presented as a means of immersion in the virtual world are such in reality. All of them must fulfil the following criteria:

- Real-time mode.
- Interactivity. It should be possible to get a response to one's own actions in the learning process.
- Three-dimensionality, allowing the realisation of a fully immersive effect.

Special VR simulators make it possible to introduce a person into a certain space that fully copies the real environment and circumstances. This is a crucial stage in the training of specialists in various spheres, allowing them to raise the level of their training. This is conditioned by the situation that we have now in the market. The consequences of errors on the part of personnel are becoming more and more fatal, and many interactions between humans and industrial equipment are realised remotely, excluding contact as such. Training on a simulator allows to develop the technology of actions in case of a real cataclysm. Training of specialists in this respect is extremely important, because it is impossible to predict the occurrence of earthquakes, which is a real misfortune for many countries.

The development of virtual reality simulators is intended for various industries and aviation. They can be made in the form of a track, platform or cube, with dimensions of 2.5 by 2.5 metres.

They are characterised by the following features:

- The person can interact with objects, they can see their arms and legs moving in different directions.
- An up-to-date digital copy.
- Modelling of various emergency and accident situations.
- Ability to integrate with real facility systems.

210

The figure shows a simulator for operational and maintenance personnel based on a virtual transformer substation. The simulator was developed to practice the professional skills of personnel during complex switching operations at 110 kV substations and allows simulating dozens of regular and abnormal situations, from workplace preparation to the need to switch equipment in emergency mode. Virtual modelling qualitatively complements the training of operating personnel and ensures effective practice of scenarios of required actions in various situations developed by the company's specialists. The use of virtual and augmented reality technology provides a high level of visibility, increases the efficiency of training, reduces its duration, and enables operational control and management of personnel in solving specific tasks.

HANG-GLIDER RIDE WITH VIRTUAL HELMET FOR STRESS REHABILITATION

The Hang Glider ride with virtual helmet is a unique combination of entertainment and therapeutic application, utilising the latest in virtual reality technology to stimulate physical and psychological activity. Optical media play a key role in this innovative direction, providing a number of benefits:

1. **Accessibility and compatibility**: Optical media facilitates distribution and access to specialised content created for the Hang Glider attraction, allowing it to be used on a variety of platforms and devices, making the technology widely available.

2. **Content security**: Sophisticated encryption and protection mechanisms used on optical discs prevent unauthorised copying and distribution of the software and multimedia content of the attraction, thus preserving the exclusivity and commercial value of the product.

3. **Support for functional and physiological rehabilitation**: Software included on optical media may contain exercises and scripts specifically designed for use in rehabilitation programmes aimed at restoring physical function and reducing anxiety in users.

4. **Easy to update and manage**: Optical discs facilitate rapid content and software updates, providing attraction operators with easy-to-use solutions for attraction maintenance and management.

5. **Scalability and diversity**: With the use of optical media, it is easy to scale and modify the experience offered by the attraction, adding new scenarios and levels of complexity, which can facilitate more personalised rehabilitation and cater to different user preferences.

The virtual helmet hang glider ride thus goes beyond mere entertainment to become a powerful tool for achieving a state of emotional balance and physical well-being, and optical media is an integral part of this process.

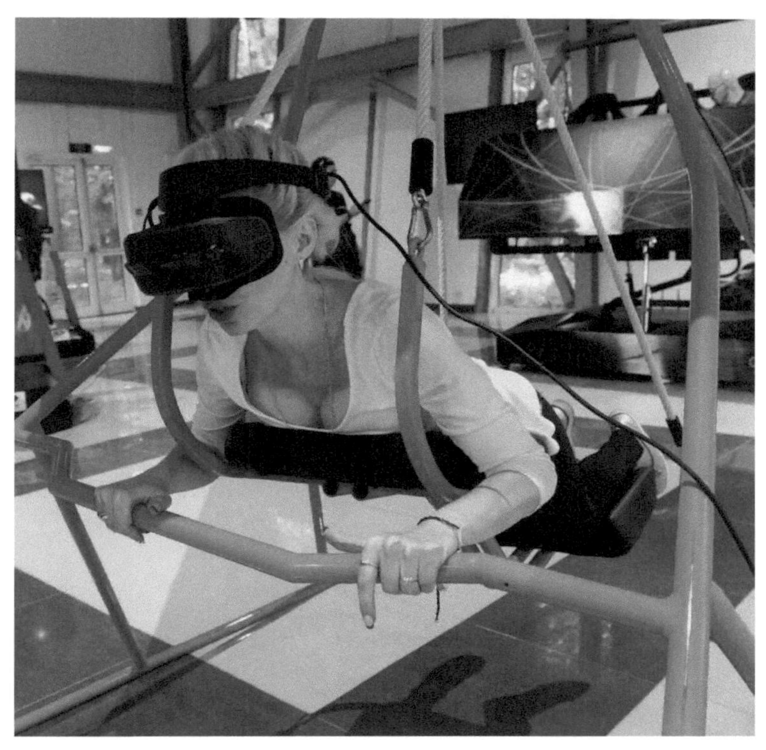

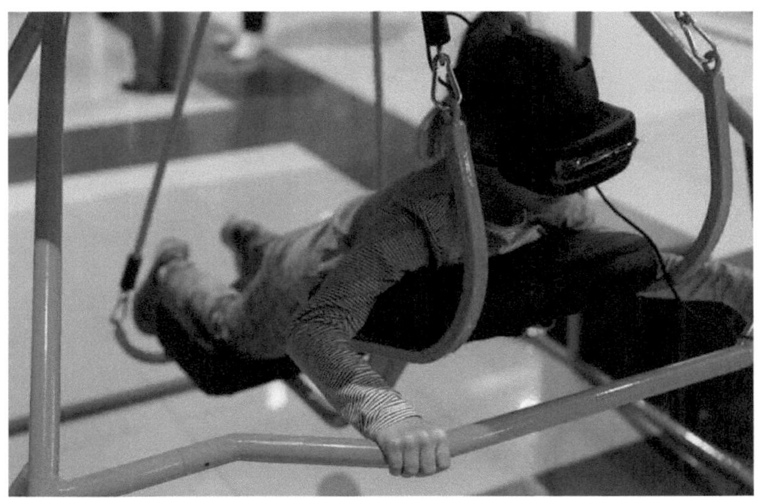

VIRTUAL REALITY VR BINOCULARS

The VR binocular is a significant tool in the development of technological aspects of functional and physiological rehabilitation thanks to its interactive optical potential. This device opens new horizons for interacting with three-dimensional objects and exploring space, offering the user a unique experience of immersion in a virtual space. This device can be used in different environments, both indoors and outdoors, thanks to its weatherproof and vandal-proof design.

The use of binoculars in rehabilitation can improve the quality and efficiency of recovery processes by offering patients stimulation of both physical and psychological activity through virtual scenarios. In particular, virtual excursions and training can be used to improve coordination, perception and other cognitive and motor functions.

In addition, VR binoculars offer new possibilities for psychological rehabilitation, using virtual environments for stress and angst reduction sessions, and as a tool for implementing therapeutic techniques for overcoming phobias and adapting to difficult social situations.

The binocular's scope of application covers museums, urban spaces, exhibitions and property demonstrations, offering volumetric visualisation and providing the user with interactive learning and exploration opportunities, making it a valuable tool in a comprehensive rehabilitation programme.

INTERACTIVE VR CINEMA ATTRACTION
VR CINEMA

VR Cinema is a combination of the latest technology applied to the viewing of modern content without a screen. VR Cinema is as easy to install as possible, uses professional components and most importantly, an open space, because now your attraction cannot go unnoticed.

VR Cinema has the 3 degrees of freedom already loved by many visitors on the upgraded hydraulic platform (forward-backward, left-right and up-down movements) and revolutionary solutions for owners of outlets in shopping centres. You no longer need to find a separate room for the attraction, because leaving it in plain sight will attract much more attention, thanks to the new design of the cinema and the much more compact and silent hydraulics that drive the platform. We also got rid of the screens and replaced them with hi-tech novelty of this year - virtual reality glasses Oculus Rift, thanks to which the visitor will be immediately transported to the 5D film from the first seconds. The changes also affected the sound, so now instead of a 5.1 home cinema system, our cinema is equipped with the most modern wireless headphones sennheiser with 3D surround sound.

The special feature of this cinema is:

- the highest quality and realism of 360 degree picture
- complete immersion in the atmosphere of the film, content
- 3 degrees of platform movement, including free fall effect

The VR Cinema interactive attraction, a cinema using virtual reality (VR) technology, offers opportunities that can be applied to functional and physiological rehabilitation:

1. Functional rehabilitation: VR Cinema allows the user to interact with a virtual environment that can help restore motor skills and coordination. Patients with motor impairments can participate in virtual exercises that mimic real physical activities, restoring limb functionality without risk of injury.

2. Physiological rehabilitation: The immersive experience that VR offers helps to reduce stress and anxiety, which is extremely useful in a rehabilitation programme. For example, VR can be used for relaxation and meditation sessions, which helps to stabilise the cardiovascular and respiratory systems.

Application:

- In medical institutions for rehabilitation after injuries and surgical interventions.
- In physiotherapy centres to restore motor function.
- In psychological practice to relieve psycho-emotional tension.

Linking to smart systems and artificial intelligence:

- Smart systems can automate the rehabilitation process by adapting VR scenarios to the patient's individual needs and progress.
- Artificial intelligence can analyse data from sensors and VR headsets to assess patient improvement, provide recommendations and personalise recovery programmes.

The use of VR in patient rehabilitation brings innovative recovery methods and provides new tools to the piggy bank of rehabilitators and therapists, allowing for a motivating and diverse environment to restore health and improve patients' quality of life.

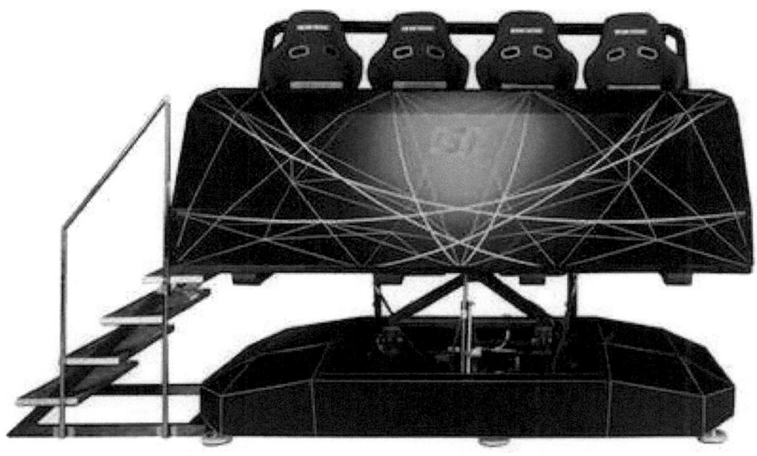

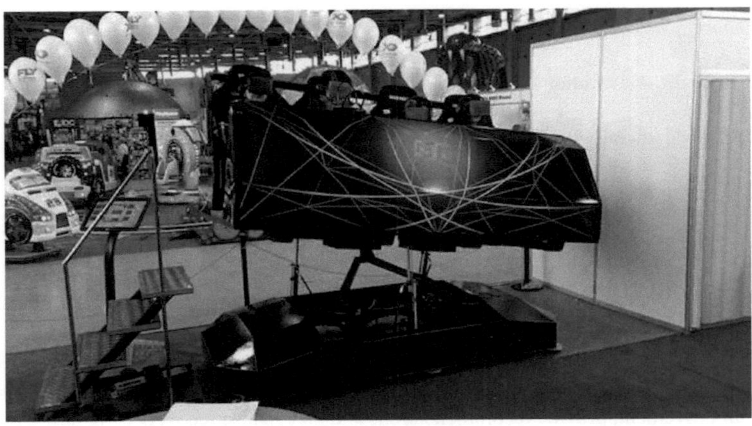

This attraction immerses visitors in fascinating worlds where you can experience movement in all directions and enjoy thrilling adventures.

The 5D cinema with VR provides an opportunity not just to watch the story, but to actively participate in what is happening. Visitors can experience moving forwards, backwards, left, right, up and down, creating a unique experience of watching short clips and all sorts of dynamic races. One of the key elements of this cinema is virtual reality, which allows viewers to immerse themselves in the atmosphere of various stories. From roller coasters to thrilling horrors and funny films, every visitor will find something to his or her liking. Here you can not only get an adrenaline rush, but also create fun and vivid memories together with your loved ones.

CAR SIMULATOR, AVIATION, HELICOPTER VR

VR simulators (auto, air, helicopter) in patient rehabilitation bring innovative recovery methods and provide rehabilitators and therapists with new tools by working as follows:

1. Improving motor skills and coordination: The use of car and flight simulators helps to improve fine and large motor skills as well as coordination. Patients learn to control their actions while driving a virtual vehicle, which is very important for those recovering from injuries or neurological diseases.

2. Increased attention and concentration: Simulator training requires constant attention and concentration, which has a positive effect on the patient's cognitive functions. These exercises are particularly effective for individuals undergoing rehabilitation after brain injury.

3. Emotional wellbeing: A source of positive emotions and stress resilience. The immersive experience of 'flying' or 'driving trials' helps to reduce stress levels and evoke positive emotions in patients.

4. Integration with artificial intelligence (AI) systems: Simulators can integrate with AI systems to tailor training programmes to each patient's individual needs. AI can also analyse a patient's progress in rehabilitation, suggesting an optimised recovery plan.

5. Use in different environments: Simulators with different degrees of freedom settings (2dof, 4dof, 6dof) allow you to simulate different environments for more targeted rehabilitation - from simple exercises for beginners to complex scenarios for advanced users.

These simulators can be used in medical rehabilitation centres, specialised injury rehabilitation clinics, centres for people with disabilities, and in private homes for supervised self-study.

These technologies greatly expand the possibilities of the rehabilitation field, making the recovery process more efficient, motivating and enjoyable for patients.

Artificial Intelligence (AI) systems are playing a key role in improving auto, air and helicopter VR simulations for patient rehabilitation purposes:

1. Personalisation of training programmes: AI can analyse a patient's skill level and needs to adjust the simulator to a specific mode of operation, helping to provide a personalised approach to rehabilitation.

2. Progress tracking and exercise adaptation: Neural networks are able to process large amounts of data in real time, allowing monitoring of the level of physical and cognitive rehabilitation progress to adjust the training programme.

3. Interactive feedback: AI can provide immediate feedback to the user on the correctness of the exercise, reinforcing the learning effect and maintaining motivation.

4. Creating immersive and realistic scenarios: Artificial intelligence helps create realistic virtual environments that stimulate natural responses and patient recovery in a safe virtual environment.

5. Predictive analytical capabilities: Sophisticated AI algorithms can predict potential difficulties in rehabilitation based on analysing patient dynamics, helping clinicians make proactive decisions to improve treatment plans.

6. Automation of simulator control: AI can automate many of the simulator's functions, making it easier for rehabilitation staff and the patient to use, especially important when working individually with each patient.

7. Simplify diagnostic work: AI-enabled simulators can help track not only the rehabilitation process, but also diagnose problems that may be hindering recovery.

Together, these features make AI an integral part of VR simulation in rehabilitation, making it more personalised, effective and engaging for patients.

In auto, air and helicopter VR simulations, artificial intelligence (AI) provides interactive feedback as follows:

1. Tracking user actions: AI using sensors and machine learning algorithms tracks the user's movements and actions in real time.

2. Evaluating performance: the AI then analyses these actions and compares them to the correct performance to assess whether the actions or exercises were performed correctly.

3. Providing feedback: Based on this analysis, AI provides interactive feedback to the user. This can be a simple success notification or the need to repeat the cycle.

TEAM GAMES IN VIRTUAL REALITY (VR)

Not only do they represent an innovative and exciting way of entertainment, but they also have the potential to be used in the field of psychological rehabilitation, also providing valuable opportunities in various fields:

Psychological rehabilitation:

- Improving social skills: VR games allow players to develop communication skills and learn to work in teams, thus improving interpersonal relationships and socialisation.
- Stress Relief: Immersion in immersive VR worlds can be an effective way to reduce stress and anxiousness, providing a safe environment to relax and unwind.
- Therapy: Using VR to engage in team games can help treat a variety of psychological conditions such as depression, PTSD and phobias by eliminating real-world risks.

Education:

- Learning: VR games can be used for educational purposes, providing an interactive and hands-on experience in teaching various subjects and skills.
- Corporate training
- Team building: VR games give corporate teams the opportunity to solve problems and develop solutions together, which strengthens team spirit and promotes leadership skills.

Health Care:

- Rehabilitation: In VR, it is possible to create simulated environments to help patients regain motor skills and cognitive function after injuries.

Family time:

- Family Activities: VR games allow the whole family to have a fun and interactive experience, strengthening family bonds.

Tourism and the hospitality industry:

- Attractions: VR games can be a unique attraction in hotels and tourist complexes, increasing visitor appeal.

Smart systems and artificial intelligence in team-based VR games can tailor the experience for the user by providing a personalised and appropriate level of difficulty, as well as collect and analyse user interaction data to improve gameplay and create deeper and more meaningful game scenarios.

PlayStation VR2
CYBER

The use of VR games in psychological rehabilitation provides a powerful tool for the treatment of depression and phobias, due to the following aspects:

1. Immersive Immersion: VR creates a fully immersive environment that helps distract from the negative thoughts and worries associated with depression. The immersive experience allows patients to feel like they are in a different place, which helps to reduce stress and anxiety levels.

2. Cognitive Behavioural Therapy (CBT): VR gaming can be integrated within CPT to treat phobias, allowing patients to face their fears in a gradual and controlled manner in a safe environment. For example, people suffering from agoraphobia can use VR to virtually visit busy places without real risk.

3. Positive Stimulation: VR gaming can stimulate the production of positive emotions such as joy and contentment, which is especially important for people suffering from depression. Positive experiences help reduce feelings of sadness and hopelessness.

4. Avatars and Virtual Interaction: In VR games, patients can use avatars to express themselves and interact with others in the virtual world. This can help overcome social isolation and develop communication skills, which often suffer in depression.

5. Control and Autonomy: VR provides the ability to control the environment and game conditions, which can be an empowerment for people whose phobias or depression are linked to feelings of helplessness and uncontrollability of the situation.

6. Progress Tracking: VR systems can track a patient's progress, providing both feedback to the patient and their therapist. This allows the rehabilitation programme to be adjusted according to the individual recovery process.

The use of virtual reality in psychological rehabilitation opens up new possibilities for effective treatment of depression and phobias, providing unique tools to overcome psychological barriers in a comfortable and controlled environment.

List of literature and patent materials used:

Annex 1 United States Patent Application	**20110033726**
Kind Code	**A1**
Jorda Sanuy; Xavier ; et al.	**February 10, 2011**

SELF-ALIGNED METAL MASK ASSEMBLY FOR SELECTIVELY DEPOSITING THIN FILMS ON MICROELECTRONIC SUBSTRATES AND DEVICES, AND METHOD OF USE

Abstract

The present disclosure relates to a self-aligned metal mask assembly for selectively depositing thin films on ***microelectronic*** substrates and devices, comprising the following parts: (a) an upper metal mask with the orifices or zones that define the patterns to be metalised, said mask having centring holes, (b) a lower metal mask with orifices of the same size and shape as the substrates or devices to be metalized, and further auxiliary holes for centring the assembly, (c) a piece or base provided with rods corresponding to the auxiliary holes, for centring the above parts, an upper piece or frame for securing and keeping the complete assembly aligned by means of screws and slight pressure. The assembly can in turn be secured to the sample-holder of the deposition machine.

Annex 2 United States Patent Application **20100154835**

Kind Code **A1**

Dimeo; Frank ; et al. **June 24, 2010**

CLEANING OF SEMICONDUCTOR PROCESSING SYSTEMS

Abstract

A method and apparatus for cleaning residue from components of semiconductor processing systems used in the fabrication of ***microelectronic*** devices. To effectively remove residue, the components are contacted with a gas-phase reactive material for sufficient time and under sufficient conditions to at least partially remove the residue. When the residue and the material from which the components are constructed are different, the gas-phase reactive material is selectively reactive with the residue and minimally reactive with the materials from which the components of the ion implanter are constructed. When the residue and the material from which the components are constructed is the same, then the gas-phase reactive material may be reactive with both the residue and the component part. Particularly preferred gas-phase reactive materials utilised comprise gaseous compounds such as XeF.sub.2, XeF.sub.4, XeF.sub.6, NF.sub.3, IF.sub.5, IF.sub.7, SF.sub.6, C.sub.2F.sub.6, F.sub.2, CF.sub.4, KrF.sub.2, Cl.sub.2, HCl, ClF.sub.3, ClO.sub.2, N.sub.2F.sub.4, N.sub.2F.sub.2, N.sub.3F, NFH.sub.2, NH.sub.2F, HOBr, Br.sub.2, C.sub.3F.sub.8, C.sub.4F.sub.8, C.sub.4F.sub.8, C.sub.5F.sub.8, CHF.sub.3, CH.sub.2F.sub.2, CH.sub.3F, COF.sub.2, HF, C.sub.2HF.sub.5, C.sub.2H.sub.2F.sub.4, C.sub.2H.sub.3F.sub.3, C.sub.2H.sub.4F.sub.2, C.sub.2H.sub.5F, C.sub.3F.sub.6, and organochlorides such as COCl.sub.2, CCl.sub.4, CHCl.sub.3, CH.sub.2Cl.sub.2 and CH.sub.3Cl.

Annex 3 United States Patent Application	**20060029727**
Kind Code	**A1**
Ivanov; Igor C.	**February 9, 2006**

Systems and methods affecting profiles of solutions dispensed across microelectronic topographies during electroless plating processes

Abstract

A method is provided which includes dispensing a deposition solution at a plurality of locations extending different distances from a centre of a ***microelectronic*** topography each at different moments in time during an electroless plating process. An electroless plating apparatus used for the method includes a substrate holder, a moveable dispense arm, and a storage medium comprising programme instructions executable by a processor for positioning the moveable dispense arm. Another method and accompanying electroless deposition chamber are configured to introduce a gas into an electroless plating chamber above a ***plate*** which is suspended above a ***microelectronic*** topography and distribute the gas to regions extending above one or more discrete portions of the ***microelectronic*** topography. An exemplary ***microelectronic*** topography resulting from the aforementioned methods and apparatuses includes a layer having distinct regions each including a comparatively different thickness and comparatively different concentrations of one of the one or more elements.

Annex 4 United States Patent Application **20070120618**

Kind Code **A1**

PEULEVEY; Nicolas ; et al. **May 31, 2007**

CIRCUIT BOARD WITH MICROELECTRONIC ELEMENTS ASSEMBLED THEREON AND METHOD FOR PRODUCING SUCH CIRCUIT BOARD

Abstract

Circuit board (1) and method of producing thereof, the circuit board having a cavity for microstrip to waveguide transition means (2) defined by a hollow space on the walls of which a protection layer (21) is provided. A ***microelectronic*** substrate (33) is placed upon an adhesive film (31) adhered onto a surface of the circuit board (1), the adhesive film being pre-cut in selected areas (32) thereof providing openings therethrough. A metal layer (5) is disposed on the resulting structure, wherein a selected part (51) of the metal layer (5) present on a surface of the microelectronics substrate 33, facing the hollow space defined by the cavity for microstrip to waveguide transition means (2) is removed.

Annex 5 United States Patent Application | **20010015176**
Kind Code | **A1**
Curtis, Gary L. ; et al. | **August 23, 2001**

Reactor for processing a microelectronic workpiece

Abstract

An apparatus for processing a ***microelectronic*** workpiece in a micro-environment is set forth. The apparatus includes a first chamber member having an interior chamber wall and a second chamber member having an interior chamber wall. The first and second chamber members are adapted for relative movement between a loading position in which the first and second chamber members are distal one another and a processing position in which the first and second chamber members are proximate one another to define a processing chamber. At least one workpiece support assembly is disposed between the first and second chamber members for supporting the ***microelectronic*** workpiece. The workpiece support assembly is operable to space the workpiece a first distance, x1, from an interior chamber wall of at least one of the first and second chamber members when the first and second chamber members are in the loading position and to space the workpiece a second distance, x2, from the interior chamber wall when the first and second chamber members are in the processing position, wherein x1>x2.

Annex 6 United States Patent Application **20030159921**

Kind Code **A1**

Harris, Randy ; et al. **August 28, 2003**

Apparatus with processing stations for manually and automatically processing microelectronic workpieces

Abstract

A method and apparatus for manually and automatically processing ***microelectronic*** workpieces. The apparatus can include a tool having a plurality of processing stations, all of which are manually accessible to a user, and an input/output station configured to support at least one ***microelectronic*** workpiece for ***automatic*** transfer to and from the processing stations. A transfer device is positioned proximate to the input/output station and the processing stations and is configured to automatically transfer ***microelectronic*** workpieces between the input/output station and the processing stations. The apparatus can be used for both manual and ***automatic*** processing of ***microelectronic*** workpieces, either sequentially or simultaneously. The processing stations can be configured to perform on the ***microelectronic*** workpiece functions such as material application, material removal, seed layer enhancement, rinsing, drying, annealing, baking, and metrology.

Annex 7 United States Patent Application	**20150177318**
Kind Code	**A1**
Di Stefano; Thomas H. ; et al.	**June 25, 2015.**

Transport Apparatus for Moving Carriers of Test Parts

Abstract

One embodiment is a transport apparatus for moving carriers of ***microelectronic*** devices along a track, the transport apparatus including: (a) a track with two rails adapted to support the carriers; (b) a trolley adapted to be transported in a direction along the track by a linear actuator; and (c) a first and a second engagement feature attached to the trolley wherein the first engagement feature is adapted to engage temporarily with a first of the carriers, and the second engagement feature is adapted to engage temporarily with a second of the carriers; wherein a predetermined movement of the trolley slidably moves the first carrier onto a test position and slidably moves the second carrier off the test position simultaneously.

Annex 8 United States Patent Application **20030217929**

Kind Code **A1**

Peace, Steven L. **November 27, 2003**

Apparatus and method for regulating fluid flows, such as flows of electrochemical processing fluids

Abstract

A method and apparatus for regulating fluid flows, such as flows of electrochemical processing fluids for processing ***microelectronic*** workpieces. The apparatus includes a valve body having an entrance port, an exit port, and a flow passage between the entrance and exit ports through which a first fluid flows. The valve body further includes a pressure chamber coupleable to a second fluid source and at least partially isolated from the flow passage. A regulator, movably disposed in the flow passage to change a flow area of the flow passage, has a first surface with a first projected area and a second surface with a second, larger, projected area, both of which are operatively coupled to the first fluid. A third surface of the regulator is operatively coupled to the second fluid. The regulator can adjust its position to maintain a constant or nearly constant first fluid flow rate as the fluid pressure at the entrance port changes.

Annex 9 United States Patent Application	**20170072081**
Kind Code	**A1**
ALVAREZ, Jr; DANIEL ; et al.	**March 16, 2017**

METHODS AND SYSTEMS FOR DELIVERING PROCESS GASES TO CRITICAL PROCESS APPLICATIONS

Abstract

Methods and delivery systems for providing a gas phase of a multi-component liquid source for delivery to a critical process application are provided. The methods include concentration of a component of the liquid source which is less volatile than water for delivery of a gas stream comprising the less volatile component to a critical process application. Critical process applications include decontamination and ***microelectronic*** processing applications.

Annex 10 United States Patent Application **20100279071**

Kind Code **A1**

Ivanov; Igor C. **November 4, 2010.**

Systems and Methods Affecting Profiles of Solutions Dispersed Across Microelectronic Topographies During Electroless Plating Processes

Abstract

A method is provided which includes dispensing a deposition solution at a plurality of locations extending different distances from a centre of a ***microelectronic*** topography each at different moments in time during an electroless plating process. An electroless plating apparatus used for the method includes a substrate holder, a movable dispense arm, and a storage medium comprising programme instructions executable by a processor for positioning the movable dispense arm. Another method and accompanying electroless deposition chamber are configured to introduce a gas into an electroless plating chamber above a ***plate*** which is suspended above a ***microelectronic*** topography and distribute the gas to regions extending above one or more discrete portions of the ***microelectronic*** topography. An exemplary ***microelectronic*** topography resulting from the aforementioned methods and apparatuses includes a layer having distinct regions each including a comparatively different thickness and comparatively different concentrations of one of the one or more elements.

LIST OF USED LITERATURE AND PATENT APPLICATIONS

Annex 1 United States Patent Application	**20180290403**
Kind Code	**A1**
Hasan; Zeaid F. ; et al.	**October 11, 2018**

Unitised Composite Structure ***Manufacturing*** System

Abstract

A method and apparatus for a composite structure ***manufacturing*** system. The system comprises a heating system and one or more ***smart*** bladders that encompass the heating system. The heating system and the one or more ***smart*** bladders form a curing tool defining a volume for the composite structure when one or more ***smart*** bladders in the one or more ***smart*** bladders is in a rigid state.

Annex 2 United States Patent Application **20180285065**

Kind Code **A1**

JEONG; Gyuhyeok **October 4, 2018**

SMART CONTROLLING DEVICE AND METHOD OF CONTROLLING THEREFOR

Abstract

The present specification relates to a ***smart*** controlling device capable of utilizing machine learning for voice recognition and a method of controlling therefor. The ***smart*** controlling device according to the present invention includes a receiver configured to receive an input including a command trigger, and a controller configured to detect one or more external display devices, select a display device of the detected one or more external display devices, cause a power status of the selected display device to be changed to a first state, and cause a response data corresponding to a first command data received after the command trigger to be output on a display of the selected display device.

Annex 3 United States Patent Application	**20180293778**
Kind Code	**A1**
Appu; Abhishek R. ; et al.	**October 11, 2018**

SMART COMPRESSION/DECOMPRESSION SCHEMES FOR EFFICIENCY AND SUPERIOR RESULTS

Abstract

A mechanism is described for facilitating ***smart*** compression/decompression schemes at computing devices. A method of embodiments, as described herein, includes unifying a first compression scheme relating to three-dimensional (3D) content and a second compression scheme relating to media content into a unified compression scheme to perform compression of one or more of the 3D content and the media content relating to a processor including a graphics processor.

Annex 4 United States Patent Application	**20180285306**
Kind Code	**A1**
ESSMANN; ROLAND ; et al.	**October 4, 2018**

INTERNET PROTOCOL (IP)-ENABLED SMART TRANSDUCER

Abstract

An Internet Protocol (IP)-enabled ***smart*** transducer includes a sensor for generating field data regarding a physical quantity associated with processing equipment or a device in an industrial processing facility, and a signal conditioning circuit for at least one of amplifying and filtering the field data to provide conditioned field data. A communications interface is coupled to the signal conditioning circuit including a processor having an associated memory and a field data to IP data conversion algorithm for generating the IP data from the conditioned field data, and a transmitter is for transmitting the IP data across an IP bus to at least one application connected to the IP bus.

Annex 5 United States Patent Application **20180272023**

Kind Code **A1**

BYSTRZYNSKI; Richard Mariusz ; et al. **September 27, 2018**

SMART OPTIC CONTROLLER FOR A HYDROXYL GENERATOR UNIT

Abstract

A hydroxyl generator unit includes an ultraviolet light source, a reaction chamber within interior space of the hydroxyl generator unit, environmental sensors and a ***smart*** optic controller coupled to environmental sensors and to the ultraviolet light source. The ***smart*** optic controller integrates environmental conditions, and in response to at least the environmental conditions, the ***smart*** optic controller generates an output signal to the ultraviolet light source to control to the hydroxyls generated by the hydroxyl generator unit. The ***smart*** optic controller includes at least one of an air flow sensor, a temperature sensor, a humidity sensor and a light sensor, to control operation of the hydroxyl generator unit. The ***smart*** optic controller includes a microcontroller that interrogates the sensors and that interfaces to external systems. The ***smart*** optic controller includes a near field communications circuit attached to a sensor communications link, and a control interface via an RS-485 connection.

Annex 6 United States Patent Application **20180272657**

Kind Code **A1**

RYU; JONGYUN ; et al. **September 27, 2018**

DISPLAY PART PROTECTOR FOR A SMART DEVICE AND METHOD OF ADHERING THE DISPLAY PART PROTECTOR TO THE SURFACE USING THE DEVICE

Abstract

Disclosed are a protector having a protecting and restoring function of a display part of a ***smart*** device and a method of adhering the display part protector. The display part protector for the ***smart*** device according to the present disclosure includes a protector part adhered to at least one of a flat display area, and a curved display area of a ***smart*** device having at least one of the flat display area and the curved display area and an adhesive layer which is formed by spreading a flow-able adhesive composition between an entire area of a lower surface of the protector part and a display area of the ***smart*** device and then curing the flowable adhesive composition and adheres the entire area of the lower surface of the protector part to the flat display area and the curved display area.

Annex 7 United States Patent Application | **20180270076**
Kind Code | **A1**
Natarajan; Sreekanth ; et al. | **September 20, 2018**

SMART NETWORKING OF TRADITIONAL APPLIANCES

Abstract

This disclosure provides systems, methods and apparatus, including computer programs encoded on computer storage media, for an appliance interface apparatus. In some implementations, the appliance interface apparatus enables network-based control of a traditional appliance. The appliance interface apparatus includes an appliance socket to connect the traditional appliance to a power source. The appliance interface apparatus includes a first network interface for establishing a first communication link with a ***smart*** appliance controller in a network. The appliance interface apparatus can establish a second communication link with a control circuit for controlling the traditional appliance. Upon receiving a command from the ***smart*** appliance controller, the appliance interface apparatus sends a signal via the second communication link to the control circuit to control an operation of the traditional appliance in accordance with the command.

Annex 8 United States Patent Application **20180270799**

Kind Code **A1**

NOH; Hoondong ; et al. **September 20, 2018**

METHOD AND APPARATUS FOR DOWNLINK CONTROL INFORMATION DESIGN FOR NETWORK COORDINATION

Abstract

A method and apparatus for downlink control information (DCI) design for network coordination (coordinated multi-point transmission (CoMP)) are provided. In addition, a method and apparatus for configuration of demodulation reference signal (DMRS) and transmission and reception of DMRS, and a method and apparatus for transmitting and receiving uplink signal according to uplink transmission scheme are provided. The disclosure relates to a communication method and system for converging a 5th-generation (5G) communication system for supporting higher data rates beyond a 4th-generation (4G) system with a technology for internet of things (IoT). The disclosure may be applied to intelligent services based on the 5G communication technology and the IoT-related technology, such as ***smart*** home, ***smart*** building, ***smart*** city, ***smart*** car, connected car, health care, digital education, ***smart*** retail, security and safety services.

Annex 9 United States Patent Application	**20180293366**
Kind Code	**A1**
Subramaniyan; Arun Karthi ; et al.	**October 11, 2018**

SYSTEMS AND METHOD FOR SECURELY SHARING AND EXECUTING DATA AND MODELS

Abstract

A simulation computer device for securely executing a model includes at least one processor in communication with at least one memory device. The simulation computer device is configured to store a ***smart*** container including a model and a usage policy. The simulation computer device is also configured to receive a plurality of inputs for the model and determine whether to validate the model based on the usage policy. The simulation computer device is further configured to execute the model with the plurality of inputs if the model has been validated. Moreover, the simulation computer device is configured to transmit at least one output.

Annex 10 United States Patent Application **20180299849**

Kind Code **A1**

MARTIN; Peter G ; et al. **October 18, 2018**

SYSTEMS AND METHODS OF HIERARCHICAL SMART ASSET CONTROL APPLICATION DEVELOPMENT AND OPTIMISATION

Abstract

Systems and methods of a Hierarchical ***Smart*** Asset Control Application development and Integrated ***Smart*** Asset Control System optimisation are disclosed. In various embodiments, the system may develop a Hierarchical Asset Control Application and corresponding control hardware requirements. This can be used to create an Integrated ***Smart*** Asset Control System in order to execute various processes for a set of equipment elements. The ***smart*** assets associated with the system may utilize intelligent agents to balance operational constraints and operational objects in order to determine real-time optimised operational parameters for a process and implement the appropriate controls to facilitate achieving the improved operational objectives.

Annex 10 **United States Patent Application** **20120029845**

Kind Code **A1**

February 2, 2012

APPARATUS AND METHOD FOR FLUID MONITORING

Abstract

According to some embodiments, an apparatus and method are provided for detecting the composition of a fluid. An alternating electromagnetic field may be applied to the fluid and distortions in the electromagnetic field are compared with predetermined, expected distortion "signatures" for particular components at particular concentrations. The presence and concentration of the components in the fluid may be detected by detecting these distortion signatures.

Annex 11 **United States Patent Application** **20130173180**

Kind Code **A1**

July 4, 2013

DETERMINATION OF ATTRIBUTES OF LIQUID SUBSTANCES

Abstract

A monitoring unit (100) that determines parameters (p1, p2) of an attribute (P) of a liquid substance flowing (F) through a dielectric conduit (110) includes plural coil members (121, 122) encircling the dielectric conduit (110) that subjects a flow of the liquid substance to plural different electromagnetic fields (B(f)), and under influence thereof measuring circuitry registers corresponding impedance measures (z(f)) of the liquid substance. A processor (130) derives the parameters (p1, p2) of the attribute (P) based on the registered impedance measures (z(f)).

Annex 12 **United States Patent Application** **20130178721**

Kind Code **A1**

July 11, 2013

VIVO DETERMINATION OF ACIDITY LEVELS

Abstract

A bolus for use in a ruminant animal's reticulum includes a cavity (100) configured to receive ruminal fluids present in the stomach. The cavity has walls (110) of a dielectric material and is encircled by a coil member (120), which is configured to subject the ruminal fluids to an electro-magnetic field. A Sensor element (310) measures the electromagnetic field's influence on the ruminal fluids and thus register an electromagnetic property representative of an acidity level of said fluids. A transmitter (410) transmits a wireless output signal (SD) reflecting the acidity measure.

Annex 13 **United States Patent** **8,820,144**

September 2, 2014

Apparatus and method for fluid monitoring

Abstract

According to some embodiments, an apparatus and method are provided for detecting the composition of a fluid. An alternating electromagnetic field may be applied to the fluid and distortions in the electromagnetic field are compared with predetermined, expected distortion "signatures" for particular components at particular concentrations. The presence and concentration of the components in the fluid may be detected by detecting these distortion signatures.

Annex 14 **United States Patent** **8,694,091**

April 8, 2014

In vivo determination of acidity levels

Abstract

A bolus for use in a ruminant animal's reticulum includes a cavity (100) configured to receive ruminal fluids present in the stomach. The cavity has walls (110) of a dielectric material and is encircled by a coil member (120), which is configured to subject the ruminal fluids to an electro-magnetic field. A Sensor element (310) measures the electromagnetic field's influence on the ruminal fluids and thus register an electromagnetic property representative of an acidity level of said fluids. A transmitter (410) transmits a wireless output signal (SD) reflecting the acidity measure.

Annex 15 **United States Patent** **9,316,605**

April 19, 2016

Determination of attributes of liquid substances

Abstract

A monitoring unit (100) that determines parameters (p1, p2) of an attribute (P) of a liquid substance flowing (F) through a dielectric conduit (110) includes plural coil members (121, 122) encircling the dielectric conduit (110) that subjects a flow of the liquid substance to plural different electromagnetic fields (B(f)), and under influence thereof measuring circuitry registers corresponding impedance measures (z(f)) of the liquid substance. A processor (130) derives the parameters (p1, p2) of the attribute (P) based on the registered impedance measures (z(f)).

Annex 16 **United States Patent** **9,194,787**

November 24, 2015

Testing apparatus for simulating stratified or dispersed flow

Abstract

A system and method for simulating stratified or dispersed flow dynamics are disclosed herein. The method includes filling an apparatus with a multi-phase mixture that includes an upper phase and a lower phase, wherein the upper phase and the lower phase are immiscible ***liquids***. The method also includes establishing contact between a bottom surface of a rotor drum and a top surface of the upper phase. The method also includes rotating the rotor drum such that the rotation of the rotor drum causes the upper phase to rotate and ultimately causes the lower phrase to also rotate. The method further includes ***monitoring*** a parameter of the multi-phase system while the rotor drum is rotating.

Annex 17 **United States Patent** **8,963,565**

February 24, 2015

Sensor for detecting liquid spilling

Abstract

The present invention concerns generally to a sensor or a sensor system for detecting spilling of aqueous ***liquids,*** for instance in confined spaces were such is critical such in an airplane. The system of present invention is an early warning system or sentinel for the prevention of corrosion by corrosive ***liquids***. Corrosion caused by corrosive ***liquids*** can rapidly change the surface properties of components in engineering structures, and that will finally endanger the functionality of structural parts. However, if ***monitoring*** technologies are in place providing continuous information on the presence of corrosive ***liquids,*** corrosion treatment and even corrosion prevention can start at a very early stage. Present invention provides such by early detection of corrosive ***liquids*** by extended sensors based on the collapse of percolation conductivity (COPC). The term collapse refers to the fact that the transition into the non-conducting state must not necessarily have the properties of a thermodynamically well-defined transition.

Printed by Books on Demand GmbH, Norderstedt / Germany